I0698547

Little Bird

and the

Walrus

A Powerful, Little Parable About How to Build a High-Performing Team and Quickly Turn Things Around - Before It's Too Late

Don Vanpool

What Others Are Saying About Don Vanpool and His Strategies

"I was genuinely engaged in the story. Not too jargony or techy. And easy to read!"

– Jennifer Burton
Executive Director of Inside Sales and Support
Kaplan North America

"The story is relatable and is a great mix of leadership, inspiration, and practical application. This book was not only enjoyable, but the lessons are valid no matter what business or industry the reader is in."

– Betsy Sullivan
Executive Director of Customer Experience
and Business Process Transformation

"Reminded me of *Who Moved My Cheese*. A clear message wrapped in a simple-to-understand story!"

– MaryAnn Camacho
CEO and Founder
Model Team Enterprises

"This book is easy to read and is highly compelling. It reminded me of the classic *The Goal,* where you have a nice mix of storytelling and powerful leadership nuggets. I wish more books were like this!"

– Stephen Kohler
Founder and CEO
Audira Labs LLC

Are You Making These Six Critical Mistakes in Your Business?

FREE eBook reveals the Top 6 Mistakes Business Owners Make...and the "Silver Bullets" you need to blast through each one and dramatically increase your cash flow, team productivity, and free time!

Get My Ebook Now at Optaprofit.com/ebook

Table of Contents

Foreword

I first met Don in October of 2017 at an entrepreneurial leadership class at our mutual alma mater, the University of Chicago Booth School of Business. We happened to sit next to one another during the class and immediately formed a bond as we discussed elements from the experience that we both found impactful. Interestingly enough, the class fueled our passion for two interrelated topics, entrepreneurship and leadership development, inspiring us to start our own businesses with the goal of supporting others.

Since that time, Don and I have become not only great friends but also strong collaborators. As we both pursued becoming certified executive coaches and starting our respective leadership development businesses (his focused on helping drive exponential business growth, and mine focused on using the lens of music to drive purpose and performance), we regularly compared our leanings along the way and found numerous ways to support and hold each other accountable.

When Don asked me to provide feedback on an early draft of *Little Bird and the Walrus*, I was not only honored but also inspired. As all great compositions do, *Little Bird and the Walrus* tells a compelling story while also infusing powerful lessons along the way that are highly applicable to everyday, real-world leadership challenges that we all face.

Among the many powerful themes included, the story covers common leadership topics, including:

- How to remain curious and practice deep active listening (in an environment when many have already made up their mind)

- How to ask deep, powerful questions to identify root-cause issues

- How to balance holding accountability on key business goals while maintaining empathy and compassion for the individual

- How to hold candid and "courageous" conversations

- How to provide clear and compelling vision and direction to your team

- How to coach and mentor team members to maximize engagement and performance

- How and when to make the hard call on tough leadership decisions that will be difficult in the short run yet powerfully beneficial in the long run

- How to create an inspired and high-performance leadership culture

One of the best parts of this book for me is that every part is based on real-life experiences that many, if not all, of us as leaders have—or will—experience at some point in our leadership journey. Highly inspiring while also extremely actionable, *Little Bird and the Walrus* is a fantastic guide for all of us to take forward in our leadership journey.

Rock on,

Stephen Kohler
Founder & CEO
Audira Labs LLC

July, 2023

Introduction

Once upon a time, in the city of Chicago, there existed a large and successful company called Chicago Tech. For years, Chicago Tech had been a prominent player in its industry, renowned for its high-quality products. However, as time passed and the business grew, several problems started to plague the company's operations, leading to a decline in its performance.

The problems began when the company's management, overwhelmed by their growing responsibilities, found it increasingly challenging to identify the root causes of their issues. This led to a lack of understanding and misalignment among various departments, resulting in a breakdown of communication and coordination. The once-effortless harmony between teams dissipated, leading to conflicts, inefficiencies, and delays in production.

Chicago Tech's management sought to find a solution to their problems and decided to hire a renowned consultant named Don Devine, known for his expertise in transforming struggling businesses into thriving enterprises. Don was known to possess exceptional analytical skills, a keen eye for identifying hidden opportunities, and a knack for inspiring positive change in organizations.

Arriving at Chicago Tech, Don keenly observed the tumultuous environment within the company. He took the time to speak individually with employees at every level, from factory workers to middle management and executives, seeking to understand their experiences, perspectives, and challenges. This approach allowed him to gather valuable insights from the people who were directly involved in the day-to-day operations.

Through his conversations and information gathering, Don learned that miscommunication, misalignment of strategic direction, and the lack of clearly defined roles and responsibilities were major obstacles within the company. Employees were often confused about their tasks, which led to errors, inefficiency, and duplicated efforts. Additionally, there was a lack of trust and collaboration between departments, causing delays in decision-making and overall productivity.

Based on his understanding of the challenges faced by Chicago Tech, Don devised a comprehensive plan that aimed to resolve the issues of miscommunication, disengagement, and inefficiency. The plan consisted of the following steps:

- Create a strategic plan for the organization: Don collaborated with the Chicago Tech leadership team to develop an evolving, written plan that set forth the vision, mission, and values of the company, long and short-range goals, and KPIs to measure progress so that the team moved together as one.

- Create a well-defined operating rhythm: Don proposed creating a well-defined communication framework that would ensure clarity and consistency while fostering collaboration across departments. Regular team meetings, updated documentation, and open channels of communication were at the core of this strategy.

- Establish clear roles and responsibilities: Don emphasized the importance of clearly defining roles and responsibilities for each employee, ensuring that everyone understood their tasks and how their work contributed to the larger goals of the company. This clarity would minimize confusion, reduce errors, and promote accountability.

- Foster a culture of collaboration: Don introduced team-building exercises, cross-departmental projects, and social events to encourage collaboration and trust among employees. Additionally, he recommended incentives and recognition programs to reward teamwork and productivity, creating a positive and unified work environment.

- Continuously improve processes: Recognizing the need for adaptation and improvement over time, Don implemented a continuous improvement program that involved regular assessments, feedback loops, and performance reviews. This allowed the company to identify bottlenecks, streamline processes, and make data-driven decisions for sustainable growth.

This is the story of that transformation.

Chapter One

Jennifer Sikorski stepped off the "L" train into the crisp fall afternoon in Chicago's Loop. She realized summer was over and the chill of winter was soon to come. She walked into a Starbucks and ordered a double espresso with hazelnut syrup. As she waited for the barista, she was excited to start another day at Chicago Tech. This was her dream job as the CEO of Chicago Tech, and it came with a big promotion and more responsibilities than she had ever had. Despite the challenges of the first few months, she had prepared for this job her entire career and felt confident she could overcome any adversity.

As she entered her office building, she felt that familiar pit in her stomach. While this was her dream job, it already felt like everything was off the rails and careening toward a deep ravine. Every day felt like the endless climb up the 110 stories of the John Hancock building she made every year as part of a fundraiser.

Jennifer walked down the hallway to her office and saw Wally, the chief engineer, standing outside the door, pacing and looking at his watch. "Good morning, Wally. How was your weekend?"

"It was okay," Wally said as he stroked his fully white, thick-as-a-brush mustache. "Have you seen the latest requests for more data from marketing? I really don't have time for all of these customer meetings, and I have more than enough information to deliver on the new product introduction in the spring."

She thought about the launch—it was already late. She knew the engineering team would never deliver a product the customer would want to purchase.

Wally was a poorly performing member of her executive team. He was an exceptional engineer and well respected. However, it was almost as if he physically dug a moat around his department and pulled up the draw bridge every morning to tinker with his engineering teams. Unless, of course, he lowers it to come to her office to complain about the other functional silos in her organization.

"Anyway, I'll address this directly with the marketing team and don't need any interference from you for now," he said abruptly.

He started to leave, but Jennifer motioned for him to stop. "Wally, this isn't about my interference. This is about you fully supporting our marketing efforts."

"I have given my support, Jennifer. I have a ten-page Power-Point I can forward on our actions to date." Wally left the room, and Jennifer could feel the tension that was left behind. She knew her biggest challenge would be the behavior of her chief engineer. Now she knew why they called him "Wally the Walrus" behind his back. He was stubborn and carried his blubber as if it were a prize.

Jennifer walked to her next meeting with her CFO and felt the all-familiar buzz on her iPhone. Lately, it felt like there was a swarm of bees in her pocket with the flow of challenges coming to her inbox. As she glanced at the latest pings, she noticed an email from her CFO to the staff with the subject line: *We will be missing our 3rd quarter cash targets.* And, of course, this would impact almost all of the plans she had for the remainder of the year and well into the next.

She turned the corner and made an abrupt turn into CFO Steve Atwater's office. "Hey Steve, I see we're having challenges with cash again this quarter. Can you explain the issues?"

"Oh hey, Jennifer—it's the same story as last quarter. We have plans that no one hits, whether it's receivables, payables, or inventory. Unfortunately, accountability has never been a strong suit in our organization."

"I can't talk now, but I have some ideas we can discuss," she said. "What I need from you is a plan on my desk by tomorrow afternoon. Missing our cash target will put all of our plans at risk. We all know that cash is king." Jennifer felt that knot in her stomach get tighter.

As she left Steve's office, she bumped into the chief marketing officer, Susan Masters. As she turned to apologize, she could see Susan was upset. "Susan, why the concerned look?" They moved into an empty office, and Susan let out a list of issues.

"Jennifer, this organization is not aligned around any of the goals we set out earlier in the year. And we are certainly not aligned around the goal of a new product launch every two months. And worse, it seems Wally is moving half the organization in the exact opposite direction. He says one thing in meetings to make everyone happy, but he does whatever he wants when everyone leaves."

Jennifer reflected on those words for a moment. "I thought the team was aligned after the offsite in Wisconsin last year. I was assured that the team had really figured out how to incorporate the company values and the one thing they would work together to accomplish. I had this conversation with Ethan before I accepted the job offer."

"No, Jennifer, just the opposite. We had some really interesting team-building events but never really went into the hard work of deep discussion and goal-setting. It seems we did a lot of touchy-feely exercises with no purpose. We need to have an offsite where we can create an action plan everyone can agree to."

"Sounds like a plan. I have to run to a meeting with the COO. I like your ideas."

"What have I gotten myself into? And how am I going to fix this?" She knew if the company failed, she would be blamed, even though the problems existed long before she arrived. These warring silos could end her promising career. Something had to be done. While she remained excited about the opportunity, she was anxious when she thought about the road ahead.

Chapter Two

The day had come Jennifer had dreaded for weeks: her quarterly performance review with Ethan Crenshaw, President of Chicago Tech's parent company, Universal Technology & Robotics. Her dream job was not going as planned, and now she was going to hear straight from the person who took a chance on her young career.

"Good morning, Jennifer, I hope you're enjoying the crisp fall air. So, how do you believe things are in your division based on your first quarter?"

"Good morning, Ethan. We definitely have a few challenges both in performance and with team dynamics. I'm working quickly to address them and feel like a plan is coming together."

"Given your performance to date, Jennifer, I would think your plan would have been forthcoming a lot sooner. I under-stand that we have more missed shipments than on-time deliveries and you are going to miss your cash targets by a significant amount. I also have not seen a plan to grow revenue or launch new products at the pace we discussed our first week. We are significantly behind in almost every conceivable way."

"Ethan, I share your concerns, and I'm working hard to address our performance issues. We have another offsite planned for next month in Door County. I'm certain we can get things back on track while getting some of our teamwork issues addressed."

Looking through glasses precariously perched on his nose, he responded in a fatherly voice. "Jennifer, I was very excited to give you this great opportunity and had all the confidence in you. But now, I'm not so certain you're up to the challenge. And the board is specifically discussing our decision and whether or not we made a mistake."

Those words felt like a dagger, and that pit in her stomach now seemed the size of a basketball. "Ethan, I'm convinced we can get this turned around, and I simply need one more quarter to get traction with the team."

"Look, Jennifer, I'm going to help you. I'm going to bring in a consulting company we've used over the last few years. They're very good at not only helping determine fixes to some of your operational issues, but they're also very good at helping the team solve difficult dynamics."

Unable to control the exasperated look on her face, Jennifer responded, "Ethan I really don't believe we need an outside person to help us fix our issues. I think they will slow us down. I really don't see how this is going to help the situation, and I can turn this around without the outside help."

"It's not really an option, Jennifer. We have ninety days to show significant improvement in performance, or the board will force my hand. This is a gift to you that I recommend you accept with open arms."

Leaning in to show at least a modicum of support, she asked, "So when will I have the opportunity to work with this consultant?"

"Their business name is Little Bird Consulting, and you can expect a visit from their lead on Monday afternoon. That will give you three or four days to prepare for the discussion and get the most out of the initial meeting."

Resigned to her fate and feeling more defeated by the minute, she responded, "Fair enough, Ethan. I'll spend some time with the staff, gather key metrics to review, and be ready Monday morning."

"Great, and why don't you make certain Wally is at this initial meeting? I know he has a vast amount of experience and has likely become your right-hand man, anyway. It would be great to bring him in from the start."

Her heart nearly stopped. This would make a bad situation worse. "I'm not so certain that is a great idea, Ethan. I have other staff members that would most likely make a bigger impact."

"No, I think Wally is best suited for this, and I have known him for years. He is a key member of our executive team. I would bring him in immediately."

Jennifer knew Wally's reputation at the corporate level was much different than the reality in her part of the business. "Sure, Ethan, I'll get with Wally immediately after our review."

"Thanks, Jennifer. I have faith in you, and I'm certain Wally will help move this along much faster and better ensure your success. And in turn our success. I'll also send him a note so he can help as much as possible."

As Jennifer left the room, she couldn't help but think the meeting could not have gone any worse. She needed to clear her head. The best way for her was to take a walk to Millennium Park and enjoy the fall air.

Chapter Three

As Jennifer left the building, she considered the conversation with Ethan. She thought that 90 to 120 days to show measurable improvement or face removal from her new position after only nine months was formidable. The challenges were both operational and with her leadership team. And she wondered what to do about Wally.

She stared into the Bean, the giant reflective mirror that attracts tourists from around the world who take selfies of themselves taking selfies of themselves.

Lost deep in thought, she saw the reflection of Stevie Bellinger, a former employee at Chicago Tech and now a senior executive in a startup headquartered in the Chicago Loop. "Hey Stevie, Great to see you again. How are things in the new company?"

"Hi, Jennifer. Honestly, I've been busy making recommended changes with a consultant we brought in to work with the team. Things were challenging at first, and I thought I had made the wrong decision. Only four months in the board was ready to make a leadership change. It was not the start I was looking for."

Jennifer listened intently and had a very knowing look on her face. "That sounds like what we're going through. I'm not totally on board though. What sort of challenges, if you don't mind me asking?"

Stevie reflected for a moment. "We seemed to always be late with shipments or new product launches. And cash was always tight to fund the business. And even worse, given the many big egos we always had challenges working together. Leadership in our new company was lacking, and we had the beginning of corporate silos before we really got off the ground."

As Jennifer listened over the sounds of the "L" Train passing, it was almost as if Stevie was describing Chicago Tech. "So how are things now, and what were some of the solutions?"

A smile creased Stevie's face as she began to answer. "I'm part of a mastermind my coach leads, and one of the CEOs mentioned a company called Little Bird Consulting. Their name reflects their approach and capabilities. The founder is able to see the big picture over the landscape and also dive deep when an opportunity to improve is pinpointed. They used a combination of consulting and coaching engagements to make rapid improvements. I was skeptical at first, but I would highly recommend them if you're looking for help."

Jennifer said, "I worked with a consultant before at my other company, and they were horrible. They didn't listen to us and came in with preconceived notions. And what was worse, they made recommendations that seemed at odds with the company values. They didn't spend any time clarifying our purpose or values or that any of their solutions were consistent. Most of our employees would see through the inconsistencies."

"Little Bird is quite the opposite, Jennifer. Not only did they tailor their approach to our exact situation, they listened intently for the first three or four weeks and conducted extensive interviews to truly understand the situation."

At first listening skeptically, Jennifer slowly started to nod as she realized Little Bird Consulting was the same company Ethan recommended. Maybe they could help. Over the horns and sirens of mid-day traffic, she asked, "So, what is the one thing that makes Little Bird different from the consulting companies you and I worked with? You know the old saying, consultants will borrow your watch to tell you what time it is."

Stevie radiated another bright smile and energetically answered. "They simply didn't assume they knew the problem upfront. They actually listened and talked to various stakeholders to understand the situation and the circumstances. This included questions related to purpose, values, and team cohesion. And as part of the solution they conducted workshops to ensure these were clear with the leadership team before we implemented solutions."

"I would add that the principal, Don Devine, is always involved. He has many years of actual business experience. It's like he can see what is coming well before it arrives. You don't get someone fresh out of business school with no practical experience."

They slowly started moving up Adams Street as Stevie continued. "Little Bird Consulting tailored their approach to fix the issue but also addressed team dynamics. They were able to deliver dramatic improvements quickly."

"Thanks, Stevie. Let's book time for a coffee on a weekend when the Bears are out of town."

As she was walking back to her office, she felt that familiar buzz again and looked down at her iPhone. It was an email from the CEO of Little Bird Consulting, Don Devine, asking for a meeting time for Monday morning. The crisp air had made her feel rejuvenated, and she realized a consulting firm with two endorsements might be just what she needed to turn things around. The conversation with Stevie convinced her of the value of Little Bird Consulting, even though she had always questioned the need for outside help. As she entered the office tower, Jim the doorman asked how her day was going. "Much better, Jim. I just had a great conversation that will help all of us."

Chapter Four

After Don sent the email to Jennifer, he checked his notes and called Ethan Crenshaw to discuss the upcoming engagement. Don knew Ethan well from their days at GE where they turned around a few failing operations. "Good morning, Ethan. It's been a while. How are you?"

"Good morning, Don. Thanks for jumping on a call this morning. I know you're busy, and I called with a short turnaround time to start the engagement with my firm. To be honest, I've been better. Our top division, Chicago Tech, is not doing well. I brought in a high-potential leader, and she's off to a slow start."

"That's what I understood from your email last week. What are some of the issues?"

"For starters, a majority of their shipments are late to customer requests, and they haven't successfully launched an NPI in months. This is causing revenue issues, and I'm very concerned about their cash position. If this doesn't get better, we may need to sell this division before we had planned."

"And what about the people in the business, Ethan? What are the dynamics?"

"Frankly, I'm having my doubts about Jennifer. She doesn't seem to work well with the head of engineering, Wally Carmichael. He has been with us for years and has been a top employee. I also don't see the team working well together. There's no indication Jennifer is able to manage the skills and personalities on her team."

"And how about the HR team? What role do they play?

"We brought in a new human resources leader, Lisa Langley. We recruited her from Silicon Valley, and she has brought in some key talent. But I believe she's frustrated and could use some help from the outside. I know this is an area Little Bird excels in. They need to implement people development processes and a system to build a cohesive team."

"We certainly can help in that area, Ethan. Tell me more about Wally."

"Sure, Don. Wally has twenty years of experience working with me specifically. Some with Chicago Tech and some with a previous business. He's an amazing engineer with multiple decades of experience. I brought him specifically to help Jennifer make the most of this part of our business."

"And what has he accomplished so far?"

"Not much, unfortunately. I can't tell if it's his working style with Jennifer or if some other dynamic is creating the issue."

Don thought to himself, *There's something familiar about that story.*

"I'll make Wally one of my first interviews when I get on site. What else, Ethan?"

"The ultimate goal here is to improve our financials, Don. Revenue and cash are big concerns. I know Little Bird is very good at seeing the big picture and then developing a clear roadmap. You break the big things down into actionable steps. And then develop the people along the way who will execute the roadmap."

Don smiled. "Yes, we have a bias for action here at Little Bird. I have a meeting with Jennifer, and this really helps me get off on the right foot. The big picture is cash and revenue. I'll get started right away, Ethan."

"Thanks, Don. I'm really looking for big things here. Your firm has always improved results."

Chapter Five

Don Devine exited the "L" train platform and walked toward the office tower while passing hurried morning commuters, excited to once again serve a new client. He was truly living his purpose and relished every opportunity to add value to a new business. He was excited to help another business stay in his hometown of Chicago.

When Don was a teenager, his father had lost a business. That altered the trajectory of Don's life and that of his family. His father's business was failing, and as the financial troubles mounted, his father decided to leave for corporate America. On an April evening after dinner, his father had informed everyone he had accepted a corporate job and the family was moving in two months.

This would be the last few weeks Don would spend living with his family. He decided to stay with friends and not make the move for his senior year. After graduation, he enlisted in the U.S. Army and began his adult life. In a matter of moments, his life had changed dramatically.

From that point on nothing remained the same. In 1984 there wasn't an internet to do research. No one had a business coach to work with small business owners, or any business owners for that matter. Of course, times changed. Now everyone uses the Internet. And business coaches are everywhere. But he was different. Because of this life event, his purpose is working with clients to improve business performance so he could have the opportunity to spare even one teenager from that fate he endured. His mission drove him.

Don spent the next few years gaining considerable leadership and operations experience. He ultimately rose to become an officer in the Army and then went to college to study finance. Following his service, he worked for the General Electric Company and earned his MBA at the University of Chicago. With all of this experience, he decided to start his own consulting company and live his purpose. He also vowed not to make the same mistakes he saw in other business consultants—he would be different and focus on serving the client.

When Don entered the building, Jennifer's assistant Mandy smiled and greeted him. "Good morning, Don. Jennifer will see you in a few moments. Please take a seat in the waiting area. Did you watch the game last night? The Bears really played well."

"I certainly did. I have season tickets, and it has been well worth the expenditure this year."

He sat down, and as he leafed through the pages of a dated *Fortune* magazine, he thought about most initial meetings and how they normally went—the usual aloofness of the client and the doubt they had that he could be of any service. He often wished they did not wait until a crisis to call him in, but invariably that was the case.

Don decided to take advantage of the extended wait and asked a few questions. "Do meetings here normally start on time?

"Not normally. Jennifer likes to have more impromptu meetings to discuss issues, so they're often created that day. Which obviously leads to some late nights and late starts. She likes to keep her calendar as free as possible, so there are not many reoccurring meetings scheduled."

"I assume that is amenable to the staff. They appreciate the impromptu nature of meetings here?"

"Some do and some do not. The real challenge is Wally, our chief engineer. He seems to dominate the time on her calendar and often the agenda of the day. It seems like he exerts undue influence."

"Confidentially," she added, "We call him 'Wally the Walrus' behind his back because he has a big, bushy mustache and moves very slowly. He holds up a lot of projects."

She looked down and noticed an instant message on her laptop screen. "Jennifer is ready to see you."

They both walked slowly toward Jennifer's office, and Don noticed the company's purpose and values on the wall. The core values mentioned customer focus, teamwork, and consistency. Based on his initial research with the company president and friends in the industry, he knew this was not the case.

He entered her office and was welcomed by Jennifer. "Please have a seat, Don. I don't have much time this morning. As you know, I have a short time frame to improve business performance. I'm not sure how an outsider can help, but I'm willing to listen. I've heard great things from colleagues, and I'm interested to see how you can have meaningful impact given our challenges."

"I hear what you're saying. Many of my new clients feel the same way. But they change their minds after they hear how I work. Let me share a few ideas with you. I don't know the impact I can have until I have better understanding of what we're trying to solve, how the team is working together, and what the data tells us about performance. That is the purpose of this initial interview. I want to get your take on the situation and better understand who in the business I should follow up with. Is that fair?"

"I suppose so."

"So, tell me, what are your top challenges?"

"We have challenges in a few areas that include fulfillment and cash. I also have issues with the executives working cohesively, there really isn't a great process to develop talent, and we don't really have a succession plan."

"Thanks, Jennifer, and what about your strategic planning process? I hear concerns about alignment and was wondering if you had something tangible to align around."

"We spent weeks creating a strategy that outlines sales, marketing, and operations strategies that I briefed to leadership last month. It's quite comprehensive."

"I'm certain it's comprehensive, Jennifer, but is the team aligned around the plan? And could I walk to the floor and ask questions to your front-line staff about the strategy?"

"Alignment seems to be an issue, and we have worked very hard to get our employees bought into the strategic direction of our leadership team. I'm not even certain the leadership team is aligned around our strategic direction. What would you suggest?"

"Maybe it would be a good idea for me to talk with three of your key staff members and ask them about clarity and alignment in the organization. And potentially three of their direct reports. I can get back to you with what I learn."

"That sounds great, actually."

"And you mentioned cash, Jennifer. What plans do you have to drive better cash performance?"

"It's probably better to discuss with our CFO, Steve Atwater. He owns our cash targets and allocates those down to each part of the business."

"Would that be part of your strategic document, Jennifer?"

"No, our strategy is more linked to concepts and less on the execution side," Jennifer answered defensively.

"Ok, I'll discuss it with the CFO as part of the interview process."

"And what about your leadership development plan and team-building efforts? How is that progressing? I notice a fair amount of anxiety about how well the team would execute a strategy even if they were aligned."

"I can direct you to our executive who runs Human Resources to answer those questions. Maybe she should be on your list of people to interview."

Don thought, *Jennifer does not think of her issues as part of her strategic planning process and mostly delegates to her staff members. And she likely does not have accountability among staff members or plans that are reviewed consistently.*

"And I believe it would be a good idea to talk with Wally. I have heard great things about him, and maybe I can get valuable insights from our conversation."

Jennifer rolled her eyes. "I'm not so certain Wally can help, but you're welcome to have that conversation."

"Great, I'll work with our assistants to schedule those meetings and will get back to you by the end of the week with my initial observations. How does that sound?"

"I can't wait to hear what you come up with. Especially with your conversation with Wally." Don detected a smirk. *What's with Wally?* he wondered.

As Don left, Jennifer thought, *Maybe working with a consultant will be a good thing after all. He asks great questions, wants to start by interviewing key staff members, and has a great reputation.*

Chapter Six

Don often started initial conversations with clients in a comfortable setting to better put them at ease. He never wanted it to feel like an interrogation or create an atmosphere that could make the client defensive. He wanted CFO Steve Atwater to feel comfortable. He needed an ally, and he knew he could solve finance problems. He needed a quick win to get the rest of the team on board. Those other problems would require time, teamwork, and buy-in from all department heads.

There was no better location for a morning meeting than Lou Mitchell's. It is located at the intersection of Jackson and Jefferson and is a historic landmark in Chicago. When Route 66 was created in 1926, Lou Mitchell's was there to greet commuters.

Don walked from the Red Line stop to the restaurant. He was in a particularly good mood given the Bears started the year with six wins in their first seven games. After twenty years on the season ticket waiting list, he finally had seats and was excited to be in the stands to cheer on his hometown team.

He entered the restaurant and saw Steve typing quickly on his iPhone. Don guessed he was responding to numerous challenges based on his previous interviews. He appeared to be agitated and likely had numerous challenges running through his inbox.

"Good morning, Steve. Thanks for taking the time this morning to discuss some of the challenges at Chicago Tech."

"Good morning, Don. Yes, my phone is buzzing like a swarm of yellow jackets. I certainly hope this isn't a long meeting. While we have our challenges to sort through, I think we have a solid plan to improve our cash flow."

"I'm certain you have a grasp on the issues and know precisely where to focus your energies. Given your MBA from the University of Chicago, I have no doubt that understanding opportunities on your financial statements is a core competency. We at Little Bird like to break this down into balance sheet opportunities and income statement opportunities. So, tell me about your approach."

"We focus on the basics of working capital. I have teams that are looking for opportunities to improve payables and receivables. I also have tasked a team to look at inventory turns and how we can improve."

He added, "And my team has a great funnel of ideas that we're ready to execute on to improve cash. We have missed our target for many months, which can be attributed to a lack of execution by the operations teams. They really don't understand the importance of cash generation."

"Tell me more about the lack of execution."

"It's simple really—we run the reports that point out very simple areas for improvement, send them to the team, and then nothing ever happens. I have even given them five or six key areas to focus on and still no movement by the team. They don't own the numbers I give them, nor do they take action on the spoon-fed ideas we hand them on a weekly basis."

"And is this the number one metric in the business? Are there competing priorities the team is working through? How does Jennifer prioritize opportunities for improvement with the leadership team?"

"We really don't have a number one metric as a team. I expect that we can work in several areas of the business and be successful. I'm not really certain why it's important for the leadership team to have singular focus. Isn't that why we hire talented executives? So, we can make an impact in all parts of our business whether that's revenue or cost?"

Don thought for a moment. The problem was the absence of a shared top goal for the team, leading to a lack of unity across all functions. He decided to discuss this with Jennifer as part of his follow-up.

"Steve, I wondered if you've looked at cash opportunities outside of working capital? We often find ways to increase cash through process work in other areas."

Looking a bit curious, Steve responded. "I'm not certain what you mean precisely. What other areas do you have in mind?"

"At Little Bird, we like to structure around balance sheet opportunities and income statement opportunities. Can I give you a few examples and send you a plan as a follow-up?"

Don continued, "Raising cash from continuing operations is a reliable and low-cost liquidity improvement opportunity. Indeed, auditing and improving operations to get the most dollars as quickly as possible should be a part of any comprehensive transformation program but is often underleveraged. Why? Because inefficiencies are not obvious and traditional management reporting tools fail to reveal what's often hidden. Complications are also a lack of operational expertise to determine what needs attention or prioritization. The good news is that there are often sizable stores of cash that businesses can tap into with the right focus, and these benefits can often be realized quickly without incurring large investment costs."

"A great example of hidden opportunities in your company for income are your call centers. In one of our key clients, the initial concern was long hold times and high abandonment rates. When we structured the conversation around process, technology, and upskilling, the answer was quite different.

Not only did we solve the hold times and abandonment rate issue with process work and basic technology, we quickly upskilled our talent to start cross-selling and upselling to existing clients."

"Other examples could be pricing improvements or new product introductions. There are many ideas across the business to improve cash. The real challenge is to get alignment with the operating teams on prioritization and then to set up structures to drive focus and execution."

Steve seemed intrigued. "And your firm can help with this? I agree change management in our company is one of the obstacles to improving not only cash but also many other challenges in our business."

"Change management is a core competency of our firm, Steve. I rely on an old equation when I engage with a client. The 'quality of an idea' times 'the acceptance of an idea' equals 'the effectiveness of the idea.' And by effectiveness, I mean does it actually get implemented and get results? They often say 'politics is the art of the possible,' and that's what we're really dealing with here."

"You know, Wally often says the same thing. I'm not certain where he picked that up, but that is something that works into many of our conversations. For all his shortcomings, Wally really does have a successful past, and I believe he genuinely cares about the organization."

Don thought, *Ok, mixed reviews on Wally.* Don asked for the check. "Thanks for your time, Steve. I'll follow up in a few days with what I hear across the business."

"Thanks, Don. I'll admit I was a bit skeptical before our meeting, but I'm much more optimistic than I was before. I'll see you at our management meeting in two weeks when you report your findings."

Chapter Seven

The next morning Don met Lisa Langley, the chief human resources officer. As she led him to her office on the 45th floor, he could look out her window to see the leaves changing in Millennium Park. He loved this time of year in Chicago and was looking forward to his meeting with Lisa.

More importantly, he saw a smile on her face that immediately reflected in his own expression. This was the first time he actually saw someone happy to meet with him.

"Good morning, Lisa. Thanks for taking time to meet with me this morning. As part of my discovery process I discuss the business with key executives, and high on my list is the human resources department. I find the technical solutions we eventually recommend aren't successful unless we work on the people dynamics."

"Good morning, Don. I've been looking forward to our meeting. I looked on your company website in preparation, and I think you could really add some value here. I really believe most of our issues are related to team dynamics."

"Tell me more."

"The biggest issue is we work in silos in our organization, and I need help breaking them down and improving collaboration. The team never really commits to making improvements, and this is the primary reason we never make progress."

"What sort of techniques or tools have you tried in the past?"

"I'm at my wits end, Don. I've tried several external coaches and consultants, making no progress. I've tried everything from rope courses to offsite wilderness experiences. All the tools and techniques you can imagine, and nothing seems to work. Everyone is guarding their own turf. I really want to hear what you have to offer."

"Tell me about your leadership development plan."

"It can seem a bit ad hoc. Given all the operational challenges, we haven't taken the time to identify and create development plans for our top leaders. I think that both frustrates the leaders looking to improve and limits the leadership ceiling in our business. Have you dealt with situations like that?"

"Yes. Here's what I've learned. There are only a few reasons that the team is not performing well. I start with understanding whether or not the people on your team have the right skills. Think of it as having the right people on the bus and in the right seats. We like to understand this first, and then we like to understand if they have the will or the right attitude to succeed. This is something we can understand and then coach those players that need to overcome. Or in some cases, remove those players from the team."

Lisa shifted a bit forward in her seat as she listened. "I like that. The combination of skill and will is simple, and I can almost place people in the categories as we speak. We do have some idea of the skillsets of most of our team, but I'm not always certain we place them in the right situations to succeed. I also don't believe we have looked closely at the will of people or the motivation to succeed in the organization. And I agree, we can coach our team members or put them in better situations to leverage their skillset."

"Great, Lisa. I like the way you're thinking about this. And tell me about your system or process to build your team. What does that look like?"

"What do you mean?"

"For example, at Little Bird, we like to access the team in a few key categories and then create a plan with the team to improve in those categories and better enable results. For example, the foundation of any team is trust. Without trust we will never have the commitment within the organization to be high performing."

"Unfortunately, we've spent little to no time thinking about team development here at Chicago Tech. It's frustrating to watch the team leave meetings with no plan or commitment to even the simplest actions to make the organization better. It's impossible to execute on new product introductions when we cannot even agree on our holiday celebration plans. It's incredibly frustrating. How do you solve this at Little Bird?"

"Ideally we have a two-day offsite with a very specific agenda. We conduct an assessment of team dynamics and then facilitate everything from building trust in the team to engaging in healthy conversations to build commitment to plans. It can be very eye-opening for the team to understand how they both view dynamics collectively and how they individually view the team."

"Another offsite? We had a consultant earlier in the year. We aren't talking about falling backward and some scary rope courses I hope."

Don chuckled. "No, nothing like that. I'm talking about building a foundation of trust in your organization and then working our way to tangible business results. We'll have very specific outcomes and metrics to ensure we are succeeding as a team. I'll admit some team members will be skeptical at first, but we always end with great results. And the team will be excited after the two-day offsite. We go offsite to rid ourselves of distractions and to focus everyone on the task at hand."

"Great, I'm all in on this one. What else do we need to discuss?"

"As part of our leadership development program I would like to conduct three-sixty evaluations of each key leader and potentially top talent one level below. Has this been done in the organization?"

"No, nothing like that. This could be incredibly powerful. Particularly with someone like Wally. He has been in the organization for years and is respected by many of the employees in his organization and by many in other departments. But I don't believe he has a realistic view of how he's perceived and the impact he has."

"Yes, I have heard about Wally, and this would be very impactful for him to see. By the way, I've heard he has a very interesting nickname."

"Oh, you mean 'Wally the Walrus.' You've really done your homework. Yes, he has a very siloed way of thinking and tends to spend hours explaining away issues rather than fixing them. He has a fixed mindset, not a growth one. And he doesn't embrace change and instead holds onto the status quo. Though, he is also appreciated by his team. I often hear they have learned more from him than from anyone else in their career. And as intimidating as he can be, he never fails to mention how many people stay on his team and never leave."

"That's consistent with what I have heard in my staff interviews. Our assessment focuses on the top two or three leadership constraints for each team member, and then we create a plan with that leader to address. All leaders have blind spots to overcome, and this process helps them make major breakthroughs."

"All of this will improve the leadership climate here at Chicago Tech, and I'm excited to begin. It seems well-thought-out and part of a complete system that'll address all elements of our team dynamics. Even Wally the Walrus could see improvement. Even though I have my doubts with him. What is the next step?"

"I will make this part of my initial assessment and discuss with Jennifer before making any recommendations. Is it ok if I follow up with questions?"

"Of course, I really am hoping to start this soon. I'm concerned with our team performance and even more with good people leaving the organization. Everyone has become frustrated, and I know some of them are actively being recruited."

"Great, I look forward to our next conversation and working together."

Chapter Eight

Later that morning, Don met Jennifer for her initial executive coaching session. He knocked on her door. "Good morning, Jennifer."

"Good morning, Don. Please come in and have a seat. I'll admit I'm a bit apprehensive about what goes on in a coaching session. So how is this going to work? What's the purpose of these coaching sessions?"

"It's perfectly normal to have those feelings when beginning executive coaching sessions, Jennifer. Our first step is to design our alliance. Every coaching relationship will be different because every person is different. The goal is to empower the coaching relationship and build trust. It all begins with one simple question: If this coaching were to have a huge impact in your life, what would it look like? For example, do you want the next promotion, or do you want to really make an impact in the life of your team?"

"And that's a question I'll answer in our next session?"

"That's correct. That will help us properly guide our discussions and create our coaching relationship. One of the unique aspects of our coaching relationship is that it is a relationship of design. You are creating me as your coach, your ally, and your champion."

"That sounds more exciting than what I was expecting, Don. I thought this was going to be more of you mansplaining to me. I'm definitely more excited now."

"Let's continue with what I call training the client. I want to discuss a few terms I'll be using. How does that sound?"

"Sounds great, Don."

"The first term is what I call the *saboteur*. You can also think of this as imposter syndrome. It's that voice in your head telling you that you cannot accomplish your goals. And the bigger the goals, the bigger the voice. When I hear the saboteur, we will call that out and ensure you recognize what is happening. I never want to coach the saboteur. I want to coach you."

"Yes, imposter syndrome sounds familiar to me. It's that voice that says I'm not up to the task or that I should've never been given this opportunity. How do I deal with that, Don?"

"That is something we will work on later. Right now, I'm making certain we're together on terms I will use. There are two other terms I want to define that are very important.

During our coaching sessions I'll often make a *request*. This word is precisely as it sounds. I will ask you to do something to further your progress and call that out as a request. You have the options to say yes, no, or give me an alternative. You always have a choice in this coaching relationship. I'll never tell you what you need to do or direct your actions."

"Other times I will issue a challenge, and I do this to get you out of your comfort zone. Accepting a challenge will immediately change your perspective and is intended to raise your level of performance. Once again you can say yes, no, or give me an alternative. Any questions so far?"

"No, I think I'm clear so far."

"Great. The last important word I'll discuss this morning is *inquiry*. This is a question to be considered over a period of time to deepen learning and provoke reflection. An example could be simply asking what you want. Sounds simple, but most people never really reflect on what they truly want. It's also the question I asked to open our session."

"That all makes perfect sense to me, Don."

"Great to hear. It's also important you understand the difference between coaching, consulting, and trusted advisor. Consulting is more about diagnosing and providing a specific solution, much like the other areas we are working on. Coaching is about the whole person. The whole person is the person who goes beyond your role as an executive. It includes health, relationships, and everyday emotions. I firmly believe you already have everything you need to succeed. It's my job to bring this out and further your progress.

"And what about a trusted advisor?"

"A trusted advisor is someone you can use as a sounding board or a thought partner. I can be all three of these roles, but most of the time I will be coaching. If we decide we need one of the other roles, we will deliberately call that out as the next action."

"Ok, that all makes sense to me, Don."

"Wonderful. The last thing I want to discuss is your three-sixty evaluation and your self-evaluation. It's very important to start with baseline information and understand what I call leadership constraints. I'll explain more as we move along."

"Thanks, Don. Yes, the team was very engaged, and I appreciate the feedback they provided. I read through everything they provided yesterday. So, how do we proceed?"

"As we discussed, I work to identify and work on leadership constraints. An example is what I call the need to nurture. Someone who scores low in the category will focus on building relationships and will earn the right to have difficult conversations. This will be an area of focus for us as it came out as a leadership constraint based on the team's feedback."

"And this seems particularly important given I'm a new leader with a new team. They're certainly not responding as I had hoped."

"That's right, Jennifer. I would ask that you think of two or three actions you can take to start building those relationships. We'll create your plan based on your ideas and some best practices I have identified."

"Sound great, Don. I'm more excited now than when we started. I can see the value in your approach."

"There are two other key leadership constraints we'll work to address. The first is overconfidence. The downside of overconfidence is the risk of intimidating others and deflecting blame on others. We will work to address this in our coaching session."

"Wow! I didn't realize that was how my coworkers viewed my leadership. I'll need to address this going forward if I'm going to help the team make the necessary changes to turn around our business."

"And the final leadership constraint we'll address is the need for control. While this is a great attribute for an executive in moderation, it can lead to power struggles with other dominant members of the team."

"Thanks, Don. All great areas to focus on over the next six months. It's great to get the view of the teams I work with and understand a few key areas that will make a big impact on the way I lead."

"That's right. This is not about big changes in behavior. You're an accomplished and successful leader, and on almost every other key leadership attribute you scored exactly where we would want an executive to score. This is about incremental changes in a few key areas that will make a material improvement in performance."

"Thanks, Don. I'm very excited to work with you and put the time into further developing my leadership skills."

"Thanks, Jennifer. In our second session we'll finish values clarification and go over the plan you create. We will also discuss your answers to the questions I asked. Great first session, Jennifer. I'm really looking forward to working with you."

Chapter Nine

Don was scheduled to meet with Chicago Tech's chief operating officer, Sherri Stillwater, in her office on the 45th floor overlooking the west side of the city. Looking out the window outside her office, he could see the United Center and the general direction of the airport. Sherri was ten minutes late, and it gave him time to think about his last meeting. Team development and leadership development would take time, and this meant his work with Sherri was even more important. Without the right operating meetings in place, they would not see any results. It was even more important to have structure in place while they worked through the people issues.

Sherri walked into her office and offered Don a seat. "I'm happy we're finally able to meet, Don. I've heard great things about your work, and we need help in many areas of our business. I'm candidly ready to try anything."

"Good morning, Sherri. I realize your calendar is full with various tactical issues you're working through, and I appreciate you taking the time to discuss your operating cadences."

"I was curious, Don. Of all things to start working on, how is this one the most important?"

"Operating cadences are often the foundation of how we run our businesses and are often overlooked. We think of them as meetings, and the last thing anyone in a company needs is another meeting."

Sherri nodded in agreement. "Precisely why I'm surprised that's the first topic we're addressing. Why would we need more meetings? I thought we would at least discuss reducing the number of meetings in our organization."

Don gave a knowing smile and continued. "It isn't the number of meetings that's important; it's how they're organized that's important. After our offsite we'll have specific metrics we're going to focus on and plans to drive improvement. What we're looking for are layered meetings to ensure daily and weekly accountability is understood."

"But we already have so many meetings. And most of them are frustrating. I would rather not spend another two hours a week discussing the same issues with our frontline teams."

"That's the real game changer, Sherri. I'm not suggesting more meetings where you spend all of your time helping direct reports and solve team problems. I'm talking about layered operating rhythms where working meetings happen before senior leadership is there to help with the truly difficult issues. That's one of the biggest issues I see at Chicago

Tech. Senior leaders spend eighty percent of their time on lower-level decisions, and the truly important twenty percent never get solved."

"So, what you're saying is we need the right kind of meetings. Not more or less. "Wally runs his department meetings in a similar way. While he doesn't work well cross-functionally, his team is good at running inter-departmental meetings."

Don noted the positive comment about Wally. "That's exactly correct. We may need more, or we may need fewer meetings. More importantly, we need the right kind of meetings with the correct people and accountability. When we know what we're trying to accomplish, we'll want working meetings to drive daily accountability with the correct participants. Then we can solve issues at the correct level, and more importantly we won't mix meeting purpose."

Sherri looked perplexed. "What do you mean by mixing purpose?"

"Some meetings need to be tactical, and some meetings need to be strategic. The purpose of some meetings is to resolve issues between teams, while some are to resolve issues across departments. Some meetings are meant to discuss strategy and not to resolve issues at all. It's important to have meetings with one clear purpose and the right level of team members to accomplish that purpose. The real frustration is when we try to do too many things in a meeting with the wrong participants."

"I get it, Don. Because we don't have a clear purpose, we not only have the wrong leaders in the meetings, we also have frustrated participants. And because of this, we are not solving the issue at the right level, if at all. And we do not have top leadership focused on the twenty percent of actions they can influence to solve the bigger issues."

Don was excited that Sherri grasped the concept so quickly. "That's exactly correct. When we come out of day two of our offsite, we will have yearly and quarterly goals with very specific metrics to measure success. We will need the right operating meetings in place to make progress. And that's where you and I will work together to put those in place."

Sherri had the look of an executive ready to get going immediately. "So, what are the next steps?"

"I'm going to complete an audit of the meetings the executive team has in place prior to the strategy session. We will compare those to the priorities we come up with as a team and start by eliminating those that do not align. We will then discuss the daily, weekly, monthly, and quarterly reviews that will make the most sense with attendees. And the team will make the final recommendations to ensure commitment."

"Love it! I sincerely look forward to working on this with Little Bird Consulting, and I'll see you at the offsite."

Don looked at his watch and realized he had 15 minutes before his meeting with Wally. He checked his messages and saw Wally canceled. No reason given. No apology. No message to reschedule.

Don thought, "This is going to be tougher than I thought."

Chapter Ten

Don's afternoon review was with the customer service vice president, Bill Evans. The fulfillment challenges in Chicago Tech were significant. Little Bird Consulting had a unique methodology to improve on-time delivery, and it was important to get alignment.

Don decided to meet in the large atrium at the Palmer House. The lobby ceiling frescos always inspired him, and he hoped they would have the same impact on Bill. Don saw Bill enter the lobby with a relaxed smile on his face and knew instantly he had made the correct decision to meet here.

"Hi Bill, thank you for meeting outside the office this afternoon. I always find that the frescos and the general atmosphere help with creativity. And with some of the challenges at Chicago Tech, I think we'll need some extra brain power. Tell me about the fulfillment challenges at the company."

"Sure, as you know, we are a business-to-consumer manufacturing company that produces various types of handheld electronics. We're lucky if we can deliver thirty-five percent on time, and we are constantly expediting product."

"Thanks for being so open and honest. And what are some of the challenges you're currently facing?"

"Well, we've been experiencing delays in our production process, and it's affecting our delivery times to customers. We also have issues with the quality of our orders coming from customers, forecasting issues, and supplier quality, and we never seem to make progress in improving our delivery performance."

"I see. Can you walk me through some of your current projects and how you prioritize your improvement efforts?"

"Absolutely, Don. We have an inventory turns team and a team that's working on forecast accuracy. They send the leadership team a report monthly on progress, and we meet with them quarterly to ensure they're on track. Most of my time, honestly, is spent expediting and working on customer complaints. It can be difficult to complete projects."

"Okay, and where are the delays happening in this project execution?"

"We mostly have issues in getting buy-in from groups outside the team and with clear project plans to keep the teams on track. And even when we do complete a project, it seems like the problem moves somewhere else in the business. There really is no clear connection between the projects we select and the on-time delivery goals we're given from the corporate office."

"I've seen this problem before. I helped another company just like yours solve this problem. Have you considered implementing lean manufacturing principles to improve project selection, execution, and overall efficiencies?"

"No, we haven't explored that yet. I've seen that term, but I don't really know what it means or how it relates to our company."

"Well, lean manufacturing involves identifying and eliminating waste in the production process and the office processes, which can lead to improved efficiency, better fulfillment, and reduced costs. We would start with a cross-functional team that would identify the entire chain of events that delivers value to the customer and then work together on a roadmap to, over time, improve the system. A cross-functional team is exactly as it sounds. We'll want the right people from marketing, sales, and operations, for example. It greatly aids with buy-in and implementation. It has an element of team building and problem-solving that's very powerful. I have seen forty-point improvements in even the most complicated supply chains."

"That sounds like it could help. How would we go about implementing lean manufacturing?"

"We start at the top of the organization. Let's get a small team to start by analyzing your current production process and identifying areas of waste. From there, we can create a plan to eliminate that waste and improve efficiency. It may also involve reorganizing your workspace and training your employees on lean principles. This will give us the data and ideas we need to get buy-in from our top leader to fully implement lean principles in all aspects of the company. And I would want you to be the executive champion."

"Okay, that makes sense. How long do you think it will take to see results?"

"It can vary, but typically it takes a few months to start seeing significant improvements. However, we can track and analyze your progress along the way and make adjustments as needed."

"All right, thank you for explaining everything to me. I think we should move forward with this. I will raise my hand to be the executive champion, and I have a team in mind to start the analysis and create the case for change. This could be a huge deal in our business."

"Great! I also believe we can couple this with the team-building systems we're creating to accelerate the effectiveness of our change ideas. 'The quality of an idea' times 'the acceptance of the idea' equals the 'effectiveness of the idea.' And lean principles will accelerate implementation and make it more permanent."

As they left, Bill took a picture of the frescos with his iPhone.

Chapter Eleven

Jennifer and Don met to discuss the strategy day planned for the offsite in Lake Geneva. This would be the culmination of the first few weeks of work and would bring everything together into one document meant to align the staff around the "big rocks" the organization needed to focus on.

Don walked into Jennifer's office and sat down while she finished a call and noticed the writing on her whiteboard. There were ideas around company purpose, values, and capabilities that needed to be built in the next year or two. He smiled as he could see some ideas he had been discussing with various staff members.

"Good morning, Don. I apologize for the delay. That was actually one of the first happy customers I have spoken with in a while. Some of the changes we made after the manufacturing kaizen your team led are having a positive effect."

"Great, Jennifer. I'm very happy to hear that. And I see you're already thinking of some of the strategic fundamentals, given the notes on your whiteboard."

"Yes, I'm starting to see some results from early implementation, and I think if we can get the strategy part correct we can at least align around a few of what you call 'big rocks.' Can you explain that again?"

"Certainly, Jennifer. In every organization there are the big priorities we have that make the biggest impact. It's the eighty-twenty concept. The key insight is that eighty percent of the impact in any organization is from twenty percent of the effort. And then we have many other smaller priorities that need to fit into any given day. Examples could be responding to customer complaints, sales funnel reviews, or employee reviews. These are all the smaller pebbles that must fit in the jar."

He continued, "If we place the small pebbles into the jar first, the big rocks will never fit. However, if we place the big rocks in the jar first, we can then get most, if not all, of the small pebbles in the jar as well. And that is what we mean by moving the big rocks first. And in the strategy session, we will boil this down to the big rocks, when they will be complete, and who owns each action."

"Got it. And so, I did what you asked and filled out the document we'll work on before the strategy session. I understand you want me to fill this in first to have my thoughts complete. And then I'll mostly observe the team as they work to complete the strategy."

"That's correct. That'll improve our changes of alignment as they will own the plan. As the leader you will get final input into the strategy. However, I find the thoughts are usually eighty percent aligned. Let's look at some of the key elements."

"Ok. I did think about our strengths and weaknesses. We have a very experienced team, and they are one of the most technically capable in the industry. But they do not work as a team, and we don't have a succession plan in place if another company poaches our talent."

"That's a great start. Now we can start to think about the short and long-term implications. The important goalposts are the long-term vision of what we want to accomplish and the quarterly goals to get there. These will be the most important elements of our plan."

"That helps clarify this initial exercise, Don. As I look at our values, I believe customer intimacy is important and product leadership. Without that we'll never survive in the marketplace we've decided to compete in."

"Ok. That will help inform some of the competencies we will need to build, which in turn will give us an idea of short-term actions. This is great, Jennifer."

"Thanks, Don. I'm starting to see how all of this fits together. And then for the 'moonshot,' I would like Chicago Tech to be the industry leader in artificial intelligence applications in our industry. And we're not even close."

"Once again. This 'moon shot' will inform what you focus on and who you need on your team. This is a great start, and I'm looking forward to the offsite."

"What are the next steps, Don?"

"Let's walk through the document in more detail and then schedule a follow-up for a final working session. I'll send a summary of the work completed, and we can ask the finance team to complete the financial projects over the next three years so we know the income and margin targets."

"Thanks, Don. Given some of the short-term wins, this could really focus the team and get us out of the trouble we are in."

"We should be proud of the early improvements in cash flow, Jennifer. Our early work has improved receivables and payables. This has provided an infusion of cash that will help build credibility with key stakeholders and, as importantly, help fund the new product ideas we're starting to generate with the marketing team."

"Yes, the reports from our finance team are showing better results than I had expected or thought possible in such a short amount of time."

Don left Jennifer's office and headed to Wally's office for his long-awaited meeting. As usual, he checked messages and saw two messages. One was from the CEO: *How's it going? Can you give me a quick call?* The other was from Wally. It simply read, *Cancel.*

After Don left Jennifer's office, she looked down at her phone to see a message from Ethan. "Hey, Jennifer, looks like cash flow is back on track. I'm really glad I hired you."

Chapter Twelve

The Friday before the offsite, Don happened to be at Chicago Tech headquarters to retrieve notes he had left in the office. He wanted to work over the weekend to summarize everything he had discussed with the team before heading to Lake Geneva on Sunday. Unfortunately, he still had not met with Wally.

As he passed by Wally's office, he noticed Wally in the corner checking emails on his phone and decided to approach him. "Good morning, Wally. Great to meet you in person finally. I did want to chat with you before the offsite to get your feedback and ideas. Do you have a moment?"

Don noticed a slight frown on Wally's face that soon disappeared. "Sure, Don. I have ten minutes before my next review. What's up?"

Don worked hard to hide his disbelief. Not a mention from Wally about the last-minute meeting cancellations and a general refusal to meet with him. "I mostly wanted to discuss the offsite and get your ideas about focus areas and initiatives. I've interviewed many of your peers on the staff, and that gives me a great context." As he sat down, he noticed several employee-of-the-month awards, his engineering degree from MIT, and pictures of his grandchildren. Wally was a Little League baseball coach based on the pictures.

"The offsite will be great, Don. I'm certain you're very good at what you do or Ethan would not have brought you in. I'll show up and engage with the team. I don't believe we need a pre-meeting. We're both very busy."

Don noticed a smirk on Wally's face and decided to press the point. "I appreciate the commitment, Wally. But it's very important for me to get pre-work completed, and you're the most seasoned and influential person on the leadership team. I'm a bit disappointed to not meet with you before the offsite."

Wally waved his hand in front of his face to signify disagreement. "It'll all be fine, and you'll get my full engagement next week. I'm certain you'll work your magic in the room and get exactly what you're looking for. No need to waste our time prior to next week."

Don realized this conversation was going nowhere. "If you change your mind, Wally, I have time this afternoon, and I'm willing to work from Chicago Tech for the remainder of the day."

"No need, Don. You have my full support, and I'll see you on Monday. Maybe send me an email with questions, and if I get time I'll send a reply."

As Don left Wally's office, he thought, *This is going to be an interesting time in Lake Geneva.*

Chapter Thirteen

The Friday before the offsite, Jennifer started making calls to ensure the team was prepared and there were no follow-up questions. She also wanted feedback on Don Devine at Little Bird Consulting.

She started with a call to Lisa Langley. "Good morning, Lisa. I'm making a few calls to ensure we're ready for next week. I know you're looking forward to the event. So how are things going?"

"Very well, Jennifer. I enjoy working with Don and his team. I'm particularly impressed with Don's professional and calm demeanor. He really commands the attention of a room and is great at managing multiple personality types. This really helps in our preparation."

"That's great to hear, Lisa. And how has that helped you specifically prepare? And do you need any help from me?"

Lisa smiled before answering. "We did a couple of prep sessions to discuss strengths and development needs of the organization with key staff members, and he was able to get the most out of the staff involved, which really helped. I'm also getting great feedback from the team that he really listens and doesn't assume he has all the right answers. He creates a space for the team to get all the issues out and be candid."

"Great to hear, Lisa. And any help needed?"

"None from me. Don has helped me prepare, and I'm really looking forward to next week."

Her next call was to Steve Atwater. "Good morning, Steve. I'm following up with key staff members before the offsite next week. Just wondering if you needed any help. I'm also looking for feedback about Don Devine."

"Good morning, Jennifer. I feel very prepared for next week. Don has helped me focus on the right data set for the strategy session and treated me like a true partner during the process. Don is a unique consultant with solid operations experience, leadership talents, and people skills. He earns my highest recommendations."

Jennifer smiled. "Great to hear, Steve. I know he is working with you to tie our financials to our strategic objectives. He is known for driving tangible results, and he's excellent at tying financials to the operations of the business."

Steve answered enthusiastically. "Yes, an area that Little Bird Consulting truly excels in. And Don truly is an operations and continuous improvement expert. He really helped me link our cash-generation capabilities with all aspects of our business operations. I'm already seeing some improvement from our collaboration."

"That is great to hear. And any help needed from me?"

"No, we have everything under control. We just need your enthusiasm and support next week. Which I know we have."

Jennifer answered enthusiastically. "Yes, you do. I cannot wait to see the results of our work next week. Have a great weekend, Steve."

Her next call was to Bill Evans, the customer service VP.

"Things are going quite well actually. Not only has Don prepared us well for the offsite, but I'm already seeing some improvements in on-time delivery from some quick wins we identified."

"That's great, Bill. And how is it working with Don?"

"It's going very well, Jennifer. Don creates a safe, yet challenging coaching environment, and our partnership has materially improved my ability to empower, develop, and challenge my team."

Jennifer's tone grew more excited. "That is great to hear, Bill. I'm receiving amazing feedback, and I can't wait to see the outcome from next week."

Jennifer decided to make one last call before heading home to start an early weekend. *Can't wait to speak to the Walrus before heading home,* she thought with a slight grimace on her face.

"Good morning, Wally. I'm calling to ask if any help is needed before the offsite next week. I would also like to get feedback on Don Devine."

Wally answered with his usual abruptness. "Good morning, Jennifer. I'm a bit busy but can provide some feedback. I'm still not certain what value Don is bringing, but I'll work with him as I have time."

Jennifer frowned. "Thanks, Wally. Any particular issue with Don or with the preparation?"

"Don tends to want more of my time than I have. He wants me to spend time with other functions in meetings. Candidly, I would think a simple interview would be enough. Even better, just send me questions in an email, and I can answer them in fifteen minutes. I don't know why he wants me to spend time collaborating with other functions before the offsite. Isn't that the purpose of the offsite?"

"Little Bird Consulting believes pre-work is ninety percent of the success, Wally. Don wants to get the most out of next week, and I agree with his approach."

Wally raised his voice noticeably. "Well, I don't agree, and I don't have the time. I'll work with everyone next week and give it my best. In the meantime, I've done all I can do.

Thanks for the opportunity, Jennifer. Will there be anything else?"

"No. I think I have all the information I need, Wally. I'll see you on Monday morning."

Jennifer decided to call Don. "Hey Don, I just had a quick conversation with Wally, and I'm a bit concerned."

Don answered with feigned surprise. "What's the issue with Wally, Jennifer? Is he not ready for our offsite next week?"

"He's simply not willing to change his style to help us succeed, Don. I'm concerned about his participation next week, and I'm even more concerned with his ability to help Chicago Tech succeed. I'm seriously considering removing him from the organization. It's a hard call but one I think may be necessary."

"I support your decision, Jennifer. I'm seeing similar issues and will back any conversation you have with Ethan."

"Thanks, Don. Let me think about this over the weekend, and I'll let you know my decision."

Jennifer hung up the phone and thought, *It's becoming obvious to me why we're in the position we're in and that we won't see dramatic improvement if Wally doesn't change. I'll speak to Ethan about this over the weekend.*

On Saturday morning, Jennifer gave Ethan a quick call. "Good morning, Ethan. Thanks for taking my call over the weekend."

"Don't mention it, Jennifer. I'm curious about the personal coaching session with Don. How is that going?"

Jennifer smiled. "Great! I feel more empowered and more willing to suggest bold moves. Which is why I'm calling. I'm considering some staff changes, and one of those people is Wally."

"I have noticed the improvement in the organization's numbers from your work with Little Bird Consulting, Jennifer. And I'm more than willing to discuss staff changes after the offsite. Let's see how Wally acts during the offsite. I want to see the direction the team is going and how they will work together. I think we owe that to everyone on your team." Ethan also thought, *I'm so glad I had the great idea to bring in Little Bird Consulting.*

"Fair enough, Ethan. We can discuss after the offsite and as part of the overall strategy."

Chapter Fourteen

The team met at the offsite location in Lake Geneva, Wisconsin, in early November. The chilly fall air had turned to clearly winter air, and the temp was well below 35 degrees. Winter typically comes early to Wisconsin and leaves late. This year would be no exception. Don pulled into the Grand Geneva parking lot and noticed Jennifer entering the building.

Don entered the main entrance and saw Jennifer in the corner working on her computer. "Good morning, Jennifer. It's great to see you early to the event. It's almost like you're excited to start our three-day team-building session." Don's smile was reciprocated.

"Good morning, Don. With all the issues still in our business I wanted to get in early to get my opening remarks prepared. Not only have we seen some early results, but I spoke with Lisa—she's very excited, and I trust her judgment."

Don decided to probe further. "What is your most pressing issue this morning?"

"Wally has missed another new product introduction tollgate review. I know our on-time delivery has improved, but, unfortunately, I don't have the new product launch ready to go for the first quarter. This means we will likely miss our revenue and cash targets."

"This is likely a great topic to cover over the next two days. A new product launch is a team sport and will probably expose issues on the team that are limiting its performance."

When everyone had filed in, Jennifer kicked off the meeting. "Good morning, team. And thank you for taking a couple of days out of your busy schedule to help us work together better as a team and to set our direction for the next twelve months. I'm very excited about some early results we are seeing, and Little Bird Consulting has a great reputation for running these sessions. I expect your full attention and participation. It will be well rewarded."

Wally was the first to speak. "I cannot imagine we'll get much out of the next three days. I've already called Ethan Crenshaw and told him I cannot believe the team is wasting its time given the challenges we are working through."

Don said, "Thanks for the feedback, Wally. I also appreciate that you shared your initial feedback with senior leadership. It's always helpful for the person who brought me in to see the stark difference in perception before and after these sessions. It shows the value Little Bird Consulting brings to the business we are working with. We will need to work together as a team to succeed. I will need your attention for the next couple of days, Wally. Anything else will be unacceptable."

The team looked at Wally with stunned surprise. No one had ever spoken to him that way before.

Don continued with the agenda. "Today will be relatively straightforward. I will walk you through the results of the team assessment, and we'll work in smaller teams to create an improvement strategy. I also have an ice breaker to start the process of building trust."

He continued, "We really are looking to improve systematically in the following categories: Trust, engagement, responsibility, accountability, and outcomes. Based on your feedback, we're red in all of these categories. While we will discuss each area, trust and team engagement will be our top areas for improvement."

Wally snapped back. "How can I trust a team that never makes any of its goals?"

"That is the point, Wally. Without trust between team members, we'll never achieve the outcomes we're looking for. I have worked with a few teams on what I call the hardware solutions, and we're seeing some results, but without a foundation of trust, this team will never consistently make its goals."

Wally chuckled. "So, a bit of chicken and egg."

"Yes, a bit of that. I will ask for cooperation and trust in the process this morning. And I will check in at lunch, and we can make any necessary adjustments based on your feedback."

Wally shook his head. "I'll give it the morning."

After Don outlined the day's agenda, he asked them to review their DISC assessment scores so they could better understand their working styles and personalities.

After discussing the assessment scores, Don kicked off the initial icebreaker. "Our first exercise is quite simple and is meant to start the process of building trust. We will combine this with your DISC scores, which will indicate the working styles of each team member. I will go around the room and ask three questions: Where were you born, what is your birth order, and what was a defining moment in your early childhood that has shaped you as a person?"

As they went around the room, Don noticed Wally waiting in an almost defiant manner. And then there was only Jennifer and Wally. "Hey Wally, why don't you give it a go?"

"Sure, why not?"

"I was born in upstate New York and was the youngest of five children. By quite a bit actually, and it almost felt like I was from a different generation than any of my brothers or sisters. Because of that my aunts showered attention on me, and I was very lucky to get whatever I wanted. Most of the time anyway. It was a great childhood. I was never really questioned about anything."

All the heads in the room nodded and had a newfound understanding of their colleague.

"I guess my defining moment was when I decided to become an engineer in high school. Realizing this was the most critical function in society and that I would make most of the rules made me feel great about the decision."

Don smiled. "Thanks, Wally. I appreciate the open discussion, and I believe it was very helpful. Jennifer, I believe you are the last team member to participate this morning."

"Sure. Much like Wally, I feel trapped in this room. I'm from Boston and was the oldest of three children. Which meant I was often the leader and made many of the decisions for the three of us. When I was ten, my father left suddenly, and we never saw him again. Fortunately for my mother, she was well-educated and was able to support our family. But she was gone much of the time, and the role of surrogate parent was left up to me. I guess this has made me very independent and able to solve problems without much interference."

And again, many of the people nodded their heads in understanding.

Don smiled knowingly. "Let's break for fifteen minutes, and then we will start with another exercise to build trust. During the break, please review your DISC results. We will discuss how to use DISC to get along better and be more productive when we come back."

The team left the room except for Lisa. "Thank you for a great start, Don. I have a better understanding of the team dynamics, and as I looked around I saw many other heads nodding. I really look forward to the rest of our day."

"You're welcome, Lisa. This will all culminate in how to get better business outcomes. Many consultants and coaches will focus on process and plans, which are all very important. I prefer to also work on what I call the software of the organization. We will have measurable, concrete plans to improve teamwork and leadership when we leave Lake Geneva tomorrow."

The start to the day was like many others he had seen. He could see the rays of hope that break through after the first couple of hours with a leadership team.

Chapter Fifteen

The team filed in, and Don could see a few smiles and hear thoughtful conversations. It would be a challenging three days, and this was a sign that the team was off to a solid start.

Don kicked off the second phase of team building. "Using DISC profiles is an effective way to team build because it helps each team member understand their own communication and behavioral styles, as well as those of their colleagues. Refer to the individualized feedback in the handouts you received prior to the offsite to help guide your discussion. I'm happy to answer any additional questions you have during the breakout sessions."

Jennifer asked, "We understand from an earlier video session how the DISC system works, Don, but how will this help build a more cohesive team?"

"By identifying individual strengths and weaknesses, team members can work together more effectively and efficiently, resulting in improved team dynamics and productivity.

Additionally, DISC profiles can help to prevent misunderstanding and conflicts by encouraging open communication and empathy."

"Can you give us an example?"

"Sure, Jennifer. Your DISC profile is a D, which is a driver. You likely have little patience for small talk, and you may grow restless listening. You are also quick to write off people whom you see as incompetent. You potentially view people with an S style, like Lisa, as indecisive and unassertive. To better communicate with her, you need to provide a safe environment, show a willingness to collaborate, and be proactive in seeking her opinion.

Lisa smiled. Several people stifled a laugh. Wally laughed out loud.

"I'm getting it now. To better communicate with the team, it's important to both understand our style and the style of the person we are working with. This way we can tailor our approach to the person to be more effective."

"Precisely. And that is the exercise we will complete this morning. I'll break you into teams, and you can discuss your styles and how to better improve communication on the team. We need to do this for two reasons. This will build trust and will be necessary to execute the strategic plan we will discuss tomorrow. Any questions?"

"None at all, Don. I'm looking forward to the exercise now that I'm clearer," said Jennifer.

"Great, let me take a few moments to describe the remainder of the offsite. We will take the rest of the day to work through the team dynamics model we started with this morning. At the end of the day, we will have three to four key actions to continue building our team to be more cohesive and productive. And by productive, I mean get results. Any questions?"

Lisa said, "Just one, Don. How will this help us get the results we will discuss tomorrow?"

"Great question, Lisa. Without the foundations, this team will not have the trust necessary to commit to plans and engage in the necessary conversation to make the plans complete. We will have a strategic plan with little chance of success."

"Great clarification, Don. And how does the rest of the session look?"

"Following today, we will work for two days on our strategic plan. We will clarify team strengths and weakness, our organizational values, our long-term plan, and the shorter-term actions to achieve our bigger goals. We also agree on the one initiative that takes precedence over all others. Think of it this way. If this team achieved only one thing this year, what would it be? We have agreement on this before we leave on day three."

Jennifer nodded to show her support. "Great process, Don. Team, this is exactly the exercise that will get us all aligned and moving forward as a team. I like to think of it as one percent strategy and ninety-nine percent alignment that will make us successful."

Everyone heard a loud *thug*.

Wally dropped his binder on the floor. Everyone looked at him.

"Oops," he said as he smiled.

Don said, "That's right, Jennifer. So, let's kick off the DISC exercise and keep forward momentum. We have an executive report to get out at the end of day three."

He thought, *I'm more confident than ever this team can succeed. Jennifer's leadership skills are improving by the day, and the team is really starting to follow her example. Except for Wally.*

Chapter Sixteen

By the end of day three, they decided on the one thing they would focus on as a group. On-time delivery was the most important initiative.

Jennifer ended the session. "Great offsite. Today we worked through our one-year goals to make our numbers and the quarterly actions we each own. And the one thing we will all focus on before any other metric in our organization is on-time delivery. This is how customers judge the success of our business."

Don added, "That's right. If we only had time to work on one thing for any given day, on-time delivery would be that one thing. It will take the entire team focused on this one metric to make improvement. This focus will also help us make progress working cross-functionally and not just in our silos."

Wally added, "And my organization and I are on board and will do everything we can to ensure we're successful as a team. Great session, Jennifer."

Don thought, "I wonder about that, Wally."

Don took Jennifer to the side and asked what she thought. Jennifer said, "I think the team is bonding nicely. I can see where this is going. Thank you very much for putting this together. This exercise was very helpful. I'm glad that we spent the time out of the office, and I know we're going to see great benefits as a result of this."

"You're welcome, Jennifer. The expectation is that we will see improvements in both performance to our plans that can be quantified and a visible improvement in team cohesion."

They then both met with Ethan via a Teams meeting to discuss the results of the session and get his feedback.

"Good afternoon, Ethan," remarked Jennifer.

"Good afternoon, Jennifer. How do you believe the session went?"

"I'm pleased with the outcome. Don did a great job leading us through the three-day session, and we have alignment on strategy and at least the start of a cohesive team."

"I agree. I like what I see coming out of the three days and look forward to progress. We certainly need to see results and very soon."

Don added, "The team took a holistic approach and identified several 'big rocks' to reach their goals. From leadership development to lean deployment in the manufacturing plants, I believe the team has the appropriate focus."

"I agree," added Ethan. "I look forward to the monthly reviews. Let me know if you need any help in the interim and best of luck."

Chapter Seventeen

Don reviewed his notes. After the offsite, he noticed that cohesion started to improve and some of the key initiatives were moving forward. There seemed to be some resistance in the organization, and Don was working to identify and reduce friction. He decided to follow up with four key executives in the organization on their "big rocks" over the course of the next two weeks before his update with the CEO.

On a very cold early December morning, Don met Lisa Langley to discuss leadership development and the team-building system they created at the offsite.

He met Lisa in her office and took a seat. "Good morning, Lisa. How are you progressing in your key areas? I know these are the building blocks for everything we're doing going forward."

"We're making progress with both of my areas of responsibility. We have created leadership development plans for three of our executives, and two of them are making progress. We realized Sherri was very focused on holding others accountable, but she had a difficult time making her teams feels comfortable. She relied too much on her ability to drive the teams and too little on her ability to understand them better as people."

"That's great, Lisa. This is often a key challenge for executives. They have a difficult time trading off accountability with caring about each team member individually. It can often be as simple as scheduling time to walk around and say hello to team members in the morning."

"Which was the opposite for Bill. He is someone who knows the names of almost everyone in customer service and cares deeply about their personal lives. He knows details that really connect him to his workforce. He also is a leader that does not always drive results and doesn't always hold those team members accountable to results. So, we are working on a very different plan with him."

"Great. And I also believe you were working with Wally. What did we identify in his three-sixty feedback and interviews?"

"And then there is Wally. He has two critical areas to work on. In his three-sixty, he came across as overly dominant with an off-the-chart resistance to change score. It is not abnormal for a person who has his years of experience to have those constraints, but he refuses to complete his development plan and never keeps our meeting to discuss his coaching plan. I'm very concerned given his importance to the success of our team."

"And as I remember he's a key contributor to the team's number one objective, on-time delivery. Given his 'big rock' of delivering new product introductions on time, he has two critical areas that he has tremendous influence on."

"That's correct. If we don't work on these personal leadership constraints, he can put our entire plan in jeopardy. We cannot go around him or ignore his behavior, and he refused to complete the plan we discussed."

"Ok. I have a meeting with him right after this and will discuss this directly with him as his assigned executive coach. Without the plan and his commitment, it'll be very difficult for me to help him and the team succeed."

"On a good note, we have all the other plans complete, and the team seems very energized. If we can get Wally on board, I think we have a great chance to turn things around here and continue to build on some early wins. I'm getting some cooperation from his operations leader, Jane Chaszar. She's a very strong leader and is on the succession plan to replace Wally if needed."

Don could see that Lisa thought some chair shifting could be in order. He had to think about the ramifications of that. "That's great, Lisa. I have scheduled with the other three top executives we discussed as part of our coaching program, and it looks like we have positive trends to build on. I'll see you after I follow up with Wally."

"Thanks, Don. See you soon."

Chapter Eighteen

Two weeks later, Don had a check-in with Ethan. He adjusted his laptop and hit the start meeting button on his calendar, and Ethan appeared on camera. "Good morning, Ethan. How are things?"

"Doing ok, Don. My final answer will be based on our conversation today. So how are things going?"

"We're seeing some improvements after the offsite in teamwork, and we have a few points of improvement in on-time delivery, which is our primary initiative. We have scheduled report-outs with you and will share those on a consistent basis."

"Thanks, Don. I realize it's too early for some of the big improvements, and I look forward to championing the work identified at the offsite. And how are things going with Jennifer?"

"At first Jennifer was reluctant, but I think she can see the value. She's certainly very coachable. The initial hesitation is perfectly normal. I would have been concerned if she was not initially skeptical of the process. I have issued a series of challenges to help her grow, and she completes work we agree needs to be done. I asked her to read a couple of strategy books, which she readily completed, and she was open to implementing new techniques I recommended. She is a great person to work with. Jennifer is very open to change and professional development."

"That is great to hear, Don. We have high hopes for Jennifer, and I believe she has a long career runway. Your coaching is exactly what we need to nurture her growth. I realize no one can get there on their own, and it's helpful to have an outside perspective to help grow our top executive talent."

"She did a great job with the offsite, and I'm seeing real improvement. It will often take six months for behavior change to take hold. I'll continue to make her development part of our time together, Ethan."

"Thanks, Don. I'm really looking forward to the next two to three months. I appreciate all the progress. Do you see any problems?"

"Well, Wally seems to be not on board. He ducks meetings. He's disruptive to his team."

"I've wondered about that," Ethan said. "He's been here a long time, and people look up to him, even if he creates friction. In the past, we tolerated that kind of behavior. I don't know if that will fly in the future. What do you think?"

"Well, that management style went out in the nineties. Today's workers won't put up with someone with that kind of attitude. Would you like me to work with him?"

"I don't think so, Don. I think we may be beyond that point. I'll give this some thought."

Chapter Nineteen

The next morning, Don looked forward to his meeting with Sherri Stillwater, the Chicago Tech chief operating officer. For the first time, the meeting was not rescheduled, nor would it start 20 minutes late. This was a good sign that the techniques they discussed were implemented and starting to work.

"Good morning, Sherri. Looks like we're running on time this morning. I can't imagine you are less busy than you were a month ago."

Sherri smiled. "No, I'm not less busy. But I'm more focused since our offsite, given the number one priority in our business is on-time delivery. Now that everyone, or at least most everyone, is rowing in the same direction, we've canceled several meetings that aren't part of our plan. As importantly, I've created the operating rhythms we discussed."

"And how is that helping?"

"It is simple. With the right players in the right meeting, we resolve issues at the appropriate level and leverage everyone's skill set. We also have clear goals and accountability for each member of our team."

Don smiled. "And how has that helped the executive team?"

Sherri nodded. "Our time is now more focused on resolving the issues our teams aren't able to resolve at their level. We no longer conduct working meetings as leaders. We use our time on the most important issues that are linked to the big rocks we agreed on at the offsite."

"And how accepting is the organization of the new meeting structure?"

Sherri smiled. "I see the frustration level at every level of the organization greatly reduced from before. Our meetings have a purpose, either tactical or strategic, and they are linked to clear goals. Productivity is at an all-time high."

"You said something earlier about almost everyone rowing in the same direction. What did you mean by that?"

Sherri frowned. "Unfortunately, Wally is not cooperating, and he is slowing progress. Without his participation we'll never reach our number one goal of eighty-five percent on time. His position in the organization is too important."

"Tell me more."

"He acts like he's part of meetings, but he is usually late and doesn't participate actively. It's also obvious his team is not using daily meetings to resolve day-to-day actions before the executive meetings. The engineering team is the only team not prepared for the weekly fulfillment review I run."

"I remember he committed to the strategic goals in our offsite. But you're telling me he's less than cooperative."

"That's right. And without his commitment, his team feels free to ignore the direction all the other teams are committed to and actively working. We cannot reach our goal without his organization's full commitment."

Don made notes on his page dedicated to Wally. He saw a clear pattern of resistance and obstruction.

"Thank you, Sherri. I'm very pleased with the progress the teams are making with the better use of layered and purposeful operating rhythms. I'll attend the Thursday fulfillment meeting to see the improvement myself. And some of the issues you discussed."

"You're welcome, Don. You're a saint."

Chapter Twenty

The next morning, Don met with Bill Evans to discuss the number one objective, on-time delivery, and the progress the team had made. He heard good things about progress thus far, with the exception of the technology team's contribution.

Don met Bill in the hallway a few steps from Bill's office. "Hey, Bill, do you mind if we have a walking meeting this morning? It helps me think better sometimes."

"Absolutely, Don. I think that's a great idea. I need a few more steps this morning anyway. I came in early due to some delivery issues caused by a new product introduction that's late to market. So, I didn't have time to exercise."

"Other than the product issues, how are things going?"

Bill smiled. "We are making great improvement. While we have considerable progress to be made, we have a fifteen-point improvement since the offsite. We simply used the framework you provided and aligned teams to specific projects. We were able to identify the projects that would deliver eighty percent of the improvement we're looking for this year."

Don returned his smile and nodded. "Great. And I assume the new operating structure we put in place is helping move the projects forward."

Bill answered enthusiastically. "The new operating rhythms are critical to the success of our projects. We have daily and weekly meetings, with the correct level of people in the meetings, solving issues before we get to the final weekly executive review. People seem to be much less frustrated."

"And how difficult was it to explain and use the framework?"

"Not difficult at all, Don. It actually makes the problem-solving much easier. We simply use the data to determine what area of the business is driving performance and align the correct team to solve the problem. Because we're focused, we not only see early progress, but also the teams are able to implement the solution faster."

"You said something about new product introductions not improving. Was that identified as a key initiative to improve delivery?"

Bill frowned. "Because we have several new product introductions this year and high demand for new products, any delay causes delivery issues that are material to our overall on-time performance. And this causes considerable customer satisfaction issues, which is reflected in our customer survey data."

"And why is the area not improving?"

"Simple. Wally says all the right things, but when it's time to focus his teams, they always seem to be working on something else. It can be very frustrating, and this has a residual impact on other teams that must work with engineering to complete their projects. His operations leader, Jane, tries to participate, but Wally diverts as soon as he becomes aware of the collaboration."

Don was about to make a snide comment before he caught himself. "I'll speak to Wally about this at a one-to-one review. Great to hear other teams are making progress and understand the approach. I've seen considerable improvement in businesses that adopt the methodology."

"Thanks, Don. I look forward to the executive review with you and the steering committee on Friday morning. The teams are justifiably proud of the positive impact they're having on our customers."

"We'll see you Friday, Bill. I have my meeting with Wally next and don't want to be late. Thanks for the time and the update. It's very encouraging."

Don checked his message and wasn't surprised to see a message from Wally. But instead of canceling, this one read, *I can meet you, but only for 15 minutes of the hour we booked.*

Don thought. "Well, let's see what we can do."

Chapter Twenty-One

Wally motioned for Don to enter his office and offered him a seat. "Sorry for the late notice, but I only have ten or fifteen minutes to touch base with you this morning. I hope that's not an issue."

"Thanks for meeting with me, Wally. I'm glad you could carve out the time to meet with me. But I have to be honest. You've been avoiding me the entire time I've been here. Did I do something to offend you? What's up? You can tell me. I can take it."

Wally pushed his glasses down so he could look over them. "Look, Don, you seem like a nice guy, but how could you possibly understand our business and all the complexity of running a tech company? I've seen guys like you come and go. It's always the same story, and nothing ever really changes. This is all a waste of my time."

"Then why did you agree to the changes in the offsite?"

"Because you will be gone soon, and I'll still be here. And I've heard all of this a hundred times. What I need to do is focus on running my time like I have for the last twenty years. I know what needs to be done, and you couldn't possibly compensate for your relative lack of experience."

"Wally, these are not my solutions. These come from your colleagues with my help as a facilitator. We need to work as a team. We can't have one person making all the decisions without the input from the rest of the organization."

Wally smirked. "Our fifteen minutes are nearly up, and I have important work to complete this morning, Don. Will there be anything else?"

"No, Wally. I think I have what I need. Thanks for the time you made available, and I'll follow up with you later."

"I doubt it."

As he walked away, Don thought, *I've seen this a hundred times, and there's only one way to resolve this for the betterment of the team.*

Chapter Twenty-Two

Don met with Ethan Crenshaw the following morning.

Ethan motioned for Don to come into his office and have a seat. "Good morning Don, I've been looking forward to our conversation. As you know, the board is looking for big improvements. I know we're making progress, but not nearly fast enough."

Don nodded. "Yes, Ethan, after the offsite we started to get traction in on-time delivery, I started key executive coaching engagements, and the team is generally working in the same direction."

"So how do we start to accelerate the changes?"

"I would start by making your engagement with the team more frequent. While the team knows you support our efforts, consistently reviewing progress with executives monthly will start to move things along more quickly. Knowing this is a top priority for you will bring along some of those who are waiting for me to end this engagement. It's always the case that a few of the leaders will wait out what they consider the initiative of the day."

Ethan frowned and nodded. "That I will do. I'll make certain my assistant protects these meetings and doesn't allow other meetings to be booked over them. I realize, as the executive champion, I haven't spent enough time with the team. What else?"

"I think it's also important you add some of our people initiatives to your quarterly strategy sessions. Discussing key executive development will not only help Jennifer, but it will also give your other senior executives an idea of best practices in her organization. I would also highlight some of the great work Jennifer's team is doing to get ideas from other executives and give some positive exposure to her staff."

"Once again, that I can do immediately. We'll add this to our quarterly agenda next month. Anything else I can do to speed progress?"

Don focused on the words we would use next. "Ethan, what I have to say next might not be a popular topic."

Ethan leaned forward, and his eyes met Don's. "What's up?"

"It's about Wally."

Ethan let out a small sigh. Don knew he could tell the story without sugarcoating it. "I know Wally has been with Chicago Tech for a long time, but I believe he's holding the team back. My sense is he is having a difficult time changing with the organization."

"Wally has been a key part of the executive team and one of our best engineers, Don. We would have a very difficult time parting with him. But this isn't the first complaint I've heard. In fact, Wally is one of the reasons I bought you in. I needed to have an outside perspective on his performance and the impact on the team. What do you suggest?"

Don let out a sigh of relief. He never liked delivering bad news. "While he has made great contributions, he is not moving ahead with the changes we agreed on in the offsite and is stopping his staff from collaborating with the other functions. We will not make the desired progress unless he changes or is removed from the organization. He either needs to be part of the team or you will need to ask Wally to leave."

"And has anyone discussed this with Wally?"

"Jennifer has worked with Wally to create his executive coaching plan. However, he has not submitted the plan and is sixty days late. I had meetings to review plans with Wally, and they were canceled on several occasions. And when we did meet he only had fifteen minutes, and it was hardly a discussion."

Ethan was visibly concerned about the implication of their discussion. "And if Wally does not work out, do we have anyone in the organization who can replace him?"

"Fortunately, succession planning and people development are part of Little Bird's core offering. I've worked with Lisa, and we identified Jane Chaszar as a potential successor. She is his right-hand person and an exceptional engineer. She has earned the respect of his organization. We decided to add her to the leadership development program, and I'm actively working with her."

Ethan seemed relieved. "Ok, my intent is to have a conversation with Wally this week and make my own judgment. I'll make a quick decision based on that conversation and any other feedback I can gather from the organization. I also have informal chains of communication to gather my own data. I do appreciate the honesty and candor. I know it is difficult to make these judgments on long-standing leaders in the organization."

"You're quite welcome, Ethan. I look forward to the actions we discussed this morning. With the changes in place and actions since the offsite, we're close to seeing very big improvements at Chicago Tech."

Don felt better after the conversation with Ethan. He called Jennifer to let her know how the conversation went.

Chapter Twenty-Three

A week later, Don had a call scheduled with Ethan and Jennifer to discuss the plan for Wally and the engineering team. He knew this would be a challenging transition for Chicago Tech but a necessary one to ensure continued improvement and the success of Jennifer. Don thought, *I see this in almost every consulting engagement, and it never gets easier.*

Don waited a few moments on Zoom, and Jennifer appeared first. "Good morning, Jennifer. How are things?"

Jennifer smiled as she was prone to do lately. "Going well, Don. Looks like we are making even more progress on our plans, and the team is working well together. With one notable exception. Which is obviously why we are here."

Don nodded and frowned for a moment. "Great to hear, and I've seen many of the reports from the team. And yes, we're here to discuss Wally's future with the organization. I do appreciate your efforts to coach Wally. Unfortunately, it isn't going nearly as well as the efforts you're making with the other executives."

"No, and now is the time for the tough conversations."

Ethan joined the conversation. "Good morning, team. I saw the latest operating reports, and things are improving. Just not in every area and not as fast as we need. Which is why we are here this morning. Jennifer, why don't you kick off our conversation with your view of Wally's performance and your recommendation."

Jennifer looked even more focused. "Sure, Ethan. Wally has not responded well to coaching sessions either from Don or from me. He doesn't work well cross-functionally and inhibits Jane from providing the leadership his organization needs. While he has had a successful career to date with Chicago Tech, he's not willing to change the way he operates within our business."

Ethan looked disappointed. "Thanks, Jennifer. And Don, what is your view?"

"I see things very similar to Jennifer. I would add that he almost never completes agreed-on actions from our coaching sessions, and the performance of his engineering team is actually deteriorating. His new product introduction launches take even longer than before, and his employee engagement scores have actually dropped. Here's another development that I just learned about. HR just told me today that we just lost Janet, our bright new start intern from Stanford. She knew more about 3D modeling and AI than anyone at the company. But she said she didn't like working with Wally.

She said he wasn't collaborative and didn't listen to her ideas. We lost a superstar."

Ethan frowned. "That's never happened in Wally's organization. He was actually sought out in our industry as a leader to work with. What is your recommendation, Jennifer?"

Jennifer leaned in and spoke firmly. "We need to offer Wally an early retirement package and promote Jane into that role. She is certainly ready and has earned the respect of the organization. She has also responded well to the coaching from Don. He helped her identify a couple of key areas to improve in, and we have seen remarkable results."

Ethan then scanned the computer screen back to Don. "Do you concur? And if so, what would be your recommendation?"

Don looked into the screen. "I agree with Jennifer's recommendation. I believe offering early retirement is the best option. The organization will see this as more than a fair outcome for Wally, and we can actually celebrate his career. Jane is one hundred percent ready for the leadership role in engineering, and I can offer transition coaching to her for ninety to one-hundred-twenty days."

Ethan turned back to Jennifer. "And have you thought of the communication outside the organization?"

Jennifer answered confidently. "Yes, Ethan. Don and I have worked on how and when to communicate and have prepared a press release. This will send a ripple through our industry given knowledge of Wally and his capabilities. Our competitors also know our failing with new products. We can use this opportunity to introduce Jane and some of her cutting-edge work. I think that will help us manage the investor community."

"Great, Jennifer. Anything else we should be thinking about, Don?"

"We also completed a risk analysis and a mitigation plan. We worked collaboratively on the coaching plan for Jane and two of her key subordinates. We also have a part for you to play, Ethan. We would like you to record a communication to all employees and then conduct several roundtables after the announcement."

Ethan smiled for the first time. "Sounds like the two of you have the plan in place to be successful. And you can count on me. I appreciate the work both of you put in. Jennifer, I have seen your growth as a leader in the last few months. I'm very proud of where you are in the organization. So, Jennifer, when do we put this in place?"

"We'll implement the plan a week from today and inform Wally this Friday late afternoon. The announcement to employees will go out next Monday at ten a.m. with a general brief to our senior staff at nine a.m."

"Thanks, Jennifer. Please text or call if you need any additional help from me. Thanks to both of you."

Chapter Twenty-Four

Spring was soon to arrive, and Don was looking forward to the Cubs home opener on Thursday. He had an appointment with Jennifer Monday morning and was moving slowly through early morning Chicago traffic on Lake Shore Drive. He thought to himself, *It'll be interesting to see how things are now that Wally has taken early retirement.*

Finally arriving, he entered the first floor of the Chicago Tech building and saw Jennifer. "Good morning, how was your weekend?"

"Great actually. We had an amazing week, and the team is really starting to work together consistently. Even the engineering team. I'm feeling more confident every day we can turn this around."

Don returned Jennifer's smile. "Great. I have follow-up meetings with staff members, and I look forward to hearing the progress."

"I think you'll be pleasantly surprised with the speed of progress over the last couple of weeks."

As they exited the elevator, he noticed a different energy in the building. Many of the team on the second floor smiled, and there was a general buzz that seemed optimistic.

Don noticed Lisa, and he asked Jennifer about the teamwork among the executive team.

Jennifer smiled. "Never been better, Don. I don't want to exaggerate, but it feels like we have made more progress in the last two weeks than we did the last four months."

"Tell me more, Jennifer."

"It's simple. The last two executive reviews were very successful. The teams were obviously working together, and we had great updates. Even where the teams did not agree, they made clear recommendations to give the executives a choice. It was obvious they had solved many of the day-to-day issues before engaging with the executives."

"And how did that help?"

Jennifer looked perplexed. "It was your doing, Don. You know how that helped. We were able to leverage the executives in the room to make the critical tradeoffs to move our big rocks forward. And because the team had worked the other ninety percent of issues, we're staying on track."

"What about leadership development? Are we making progress?"

Jennifer beamed. "We're making exceptional progress. And it's much easier now that I'm not asking the most senior team member for his plan every day. Wally's replacement has already completed hers."

"That's great. Do you mind walking me around, Jennifer?"

"Not at all."

As they walked around the office, Don saw signs that teams were working better together. Employees at every level discussed their link to the big rock and how they were contributing. Most everyone could recite the number one initiative and why it was important.

As he was ready to leave the building, he thanked Jennifer. "I appreciate you taking the time this morning, Jennifer. The team looks more engaged than ever before."

"It's just like I said, Don. We just needed to get everyone moving in the same direction and have a candid conversation with those who were not working with the team."

Don left the building and thought to himself, *Amazing how addition by subtraction always works.*

Chapter Twenty-Five

Jennifer met with Stevie over coffee to discuss the changes at Chicago Tech. "Thanks for taking time this morning, Stevie."

"Of course. I'm curious about progress in your business and your work with Little Bird."

"Initially I was busy putting out fires and resisting the idea of working with an outside consultant."

Steve nodded. "That's where I was initially, Jennifer. And things worked out well for me."

"Me too. After ten months of working with Don at Little Bird, I gained a new confidence in my abilities thanks to his coaching. And the company's numbers were up because of his suggestions and the improvement in teamwork at Chicago Tech."

"And I'm certain your energy, curiosity, and execution skills also made a huge difference."

"I would certainly like to think so, Stevie," Jennifer said with a smile.

"I really am happy for you, Jennifer. Do you have any follow-up with Don?"

"We certainly do. As part of his service, he'll continue to coach some of our key executives for a few more months, and we have several follow-up engagements to continue building our operational capabilities. I'm really looking forward to the continued collaboration."

"Well, I'm happy for you, Jennifer. If you need anything from me, please let me know."

Jennifer picked up the check, and as she was leaving the coffee shop, she received a text from Ethan that said, *Come see me. It's time for your annual review. I think you'll be very happy with what I'm going to propose for your next career move.*

Epilogue

Don worked tirelessly alongside Chicago Tech's management and employees to implement his strategies. He facilitated workshops, conducted training sessions, and personally mentored key personnel to help them adapt to the proposed changes.

Gradually, the effects of Don's interventions became apparent. The communication gaps closed, fostering a greater sense of purpose and collaboration. With clearly defined roles, employees became more efficient and effective, their true potential finally being realized. Trust among teams began to grow, and innovative ideas flourished as individuals worked together toward common goals. And they moved the big rocks, which led to on-time delivery to customers exceeding 95 percent and made them the best in their industry in customer satisfaction.

And Don has coached businesses through many of these turnarounds. All evidence of the importance of bringing in the right consultant who will lead the business through the discovery of opportunities by including the right stakeholders from the beginning. The roadmap includes alignment around strategic goals, focusing on tangible business results, building communication strategies into daily operations, improving team cohesion, and coaching key executives.

Acknowledgements

I am grateful to those who have made this book possible both in my personal and professional life. The people I have worked with have taught me much and brought joy to my life. I am grateful to the very many people who have influenced the writing of this book. Tony Sikorski, MaryAnn Camacho, Stephen Kohler, Jennifer Burton, Betsy Sullivan, Jim Ambrose, Catherine Estrampes, Larissa Chaikowsky, Todd Bertulis, Jeff Davey, Eric Dombach, Karl Bryan, Pam Gieg, Dave Hammers, Darrell Jack, Mollie Knopka, Jon Leverance, Fabien Metivet, Devina Mistry, Jay Nunes, Andy Penca, Ray Roberge, Skip Gallo, Jane Chaszar, Andy Temte, Jesse Vanpool, Carey Vollmers, Steve Spivey, Sam Allison, Dan Janal, Christopher Foye, Nancy Conger, Ralph Seely, Will Bedingfield, Nick Oak, Mike Peyer, Deb Hemstock, and the Walrus.

About Don Vanpool

Don Vanpool is a seasoned executive and coach with over three decades of experience in corporate leadership and strategic consulting within the Fortune 500 landscape.

He has launched and led three successful businesses, held a prestigious executive role at General Electric (GE), engaged in private equity ventures, and proudly served as a military officer in the esteemed 82nd Airborne Division.

Currently, as the CEO of OptaProfit, Don steers the business coaching firm toward a singular mission: driving profitable growth, building teams, cultivating executive talent, and optimizing strategic operations. His adept coaching and mentorship have empowered countless business leaders, fostering their professional development while achieving tangible business improvements.

Don's impact on the business world extends across diverse industries, from spearheading transformative initiatives in manufacturing businesses of various sizes to navigating the complex terrain of banking and pioneering innovative solutions in digital education.

He earned his MBA from the prestigious University of Chicago and earned the esteemed Six Sigma Master Black Belt certification.

He can be reached at donvanpool@optaprofit.com.